GPS Civil Monitoring Performance Specification
DOT-VNTSC-FAA-09-08
April 30, 2009

DEPARTMENT OF TRANSPORTATION

GLOBAL POSITIONING SYSTEM (GPS) CIVIL MONITORING PERFORMANCE SPECIFICATION

Thomas J. Nagle, DOT/FAA
Program Manager, Civil Applications
Global Positioning Systems Wing/GPC
483 North Aviation Blvd.
El Segundo, CA 90245-2808

DISTRIBUTION STATEMENT A. Approved for public release; distribution is unlimited once approved.

THIS PAGE INTENTIONALLY LEFT BLANK.

Executive Summary

This Civil Monitoring Performance Specification (CMPS) is published and maintained at the direction of the Program Manager for Civil Applications, Global Positioning Systems Wing (GPSW). The purpose of this document is to provide a comprehensive compilation of requirements for monitoring the GPS civil service and signals based on top level requirements to monitor all signals all the time. Upon approval this CMPS will be used by the GPS community to determine the adequacy of civil monitoring and provide focus for any needed monitoring improvements.

This CMPS defines a set of metrics for assessing GPS performance against standards and commitments defined in official U.S. Government documents such as the Standard Positioning Service Performance Standard, the Navstar GPS Space Segment/Navigation User Interfaces (IS-GPS-200), Navstar GPS Space Segment/User Segment L5 Interfaces (IS-GPS-705), and Navstar GPS Space Segment/User Segment L1C Interfaces (IS-GPS-800). This CMPS will be revised to track changes in these key reference documents. The implementation of a system that satisfies these requirements will allow operations as well as users to verify that civil GPS performance standards and commitments are achieved. To the extent practicable, each metric defined is traceable to one or more specifications or commitments of performance. In cases where the metric is an indirect measurement of performance, the connection between the metric and the standard is explained and the threshold and/or goal necessary to achieve acceptable performance provided.

This document also defines the scope and range of monitoring needs not directly traceable to the key reference documents but expected by civil users. These needs include the ability of the service to detect defects in signal and data, the rapid report of anomalous service behavior to satellite operations for resolution, and notification to users of the causes and effects of such anomalies for their various service types (e.g., positioning, timing, and navigation). This CMPS also addresses the need for archives of key data and events to support future improvements in GPS service and to respond to external queries about actual GPS service levels.

This CMPS addresses the current L1 C/A signal and the GPS Standard Positioning Service (SPS) provided via that signal. It also includes the planned L1C, L2C, and L5 signals along with semi-codeless use of the GPS signals.

This performance specification is *not* intended to state how civil monitoring will be implemented nor does it address the monitoring system architecture. The purpose of this CMPS is to provide the current requirements for monitoring of the civil service and signals for use by the U.S. Government in planning GPS development efforts. As a result, many of the requirements contained in this CMPS may be incorporated into the next generation operational control system (OCX), while other requirements may be allocated to other government entities for implementation.

GPS Civil Monitoring Performance Specification

Table of Contents

1 SCOPE 1
   1.1 Scope 1
   1.2 Background 1
   1.3 Document Description 2
2 APPLICABLE DOCUMENTS 5
   2.1 General 5
   2.2 Government documents 5
      2.2.1 Specifications, standards, and handbooks 5
      2.2.2 Other Government documents, drawings, and publications 5
   2.3 Non-Government documents 6
3 REQUIREMENTS 7
   3.1 System Performance Monitoring Requirements 7
      3.1.1 Verification of Constellation Management Standards 7
      3.1.2 Verification of Signal in Space Coverage Standards 7
      3.1.3 Verification of Signal in Space Accuracy Standards 7
      3.1.4 Verification of Signal in Space Reliability Standards 8
      3.1.5 Verification of Signal in Space Continuity Standards 8
      3.1.6 Verification of Signal in Space Availability Standards 9
      3.1.7 Verification of Position/Time Domain Availability 9
      3.1.8 Verification of Position/Time Domain Accuracy 10
   3.2 Civil Signal Monitoring Requirements 10
      3.2.1 Verification of Civil Ranging Codes 10
      3.2.2 Civil Signal Quality Monitoring 11
      3.2.3 Verification of Signal Characteristics for Semi-Codeless Tracking 13
      3.2.4 Verification of Navigation Message 13
   3.3 Verification of GPS III Integrity 18
   3.4 Non-Broadcast Data Monitoring Requirements 18
   3.5 Reporting and Notification requirements 18
   3.6 Analysis and Data Archiving Requirements 19
   3.7 Infrastructure Requirements 20
   3.8 Operations Integration Requirements 20
4 Partitioning of Requirements 21
5 NOTES 22
   5.1 Additional References 22
   5.2 GPS Civil Monitoring Service Use Cases 22
      5.2.1 GPS Operational Command and Control 22
      5.2.2 GPS Service Standard Adherence 23
      5.2.3 GPS SPS and IS Compliance 23
      5.2.4 Situational Awareness 24
      5.2.5 Past Assessment 25
      5.2.6 Infrastructure 25
      5.2.7 Power Level Assessment 26
   5.3 Allocation of Requirements to Use Cases 27
   5.4 Clarifications and Algorithms 28

- 5.4.1 Verification of Absolute Power 28
- 5.4.2 Received Carrier to Noise 28
- 5.4.3 Code-Carrier Divergence and Code-Carrier Divergence Failure 28
- 5.4.4 Signal Distortion 29
- 5.4.5 Carrier Phase and Bit Monitoring 30
- 5.4.6 Assessment of DOP Availability 31
- 5.4.7 Position/Time Domain Accuracy 35
- 5.5 Definitions 39
- 5.6 Abbreviations and Acronyms 40

THIS PAGE INTENTIONALLY LEFT BLANK.

1 SCOPE

1.1 SCOPE

This Civil Monitoring Performance Specification (CMPS) establishes the performance and verification requirements for monitoring the GPS civil service and signals.

1.2 BACKGROUND

The Global Positioning System (GPS) was not initially envisioned as a worldwide civil utility. It was created as a military navigation system, and was designed to meet warfighter needs. As a result, the service provider monitors only the Precise Positioning Service (PPS). The service provider has never continuously monitored the Standard Positioning Service (SPS) signal. Indeed, the Coarse/Acquisition (C/A) code is monitored by the service provider for a brief time after the satellite rises above the horizon of each GPS monitor station and while initial acquisition is in progress.

The U.S. Government redefined the mission of GPS to include international civil users in 1983. In that year, a Korean airliner drifted off course and was shot down by the Soviet Union when it allegedly flew over restricted airspace. Soon after, President Reagan issued a statement saying that GPS would be available for international civil use. This policy was formalized in 1996 when President Clinton declared through a Presidential Decision Directive ("U. S. Global Positioning System Policy") that GPS was a civil and military service, to be provided on a continuous worldwide basis free of direct user fees. This policy was later codified in United States Code (USC), Title 10, Section 2281.

Until May 2000 the GPS service provider intentionally degraded the SPS signal in an effort to deny accurate positioning service to U.S. military adversaries. Elimination of the intentional degradation of SPS stimulated increased use and dependence on the SPS signal. The addition of two additional civil signals (L2C and L5) will result in further increase civil reliance on GPS. Eventually, the addition of a new interoperable signal called L1C with the EU Galileo Program will facilitate the process of integrating GPS into the international GNSS community. The increased size and international diversity of the user community argues for greater importance of monitoring, while the increased performance, especially accuracy, of the service along with these additional civil signals implies that accomplishment of the necessary monitoring will be challenging.

Through the GPS SPS Performance Standard (SPS PS), the U.S. Government establishes a basis for the level of service provided for civil users. The document states that it "defines the levels of Signal In Space (SIS) performance to be provided by the USG to the SPS user community." This CMPS then provides the desired monitoring capabilities that will serve to verify the SPS PS performance.

The military/government use of GPS represents but a small fraction of the total economic impact of GPS; civil applications represent the vast majority and have an estimated dollar amount approaching $16 billion for 2003 [Department of Commerce brief, February 2001]. GPS serves many expanding and emerging commercial markets: aviation, precision farming, survey and mapping, maritime, scientific, timing, embedded wireless, space navigation, and terrestrial navigation, to name a few. Additionally, GPS-based safety, security, navigation and information systems will play a major role in the implementation of the Intelligent Transportation System (ITS) and other telematic systems of the future. GPS's growing integration in safety-of-life applications compels the basic SPS service to be adequately monitored to ensure compliance with stated performance standards.

Access to measurement, notification, and status data is important to users and agencies associated with GPS. For this reason, the U.S. Government operates several data services, including the publication of Notice Advisories to Navstar Users (NANUs), almanacs, and performance assessments. This data is currently disseminated to users through websites and by other means established by the FAA, the U.S. Coast Guard, and the U.S. Air Force. In the future, access to such data will be enhanced by the implementation of the Global Information Grid (GIG). As the GIG technology becomes available, reports and notifications resulting from the monitoring activity will be included in the data stream that is sent to users worldwide.

In addition to the dissemination of monitoring data, the future could include the use of measurement data from nontraditional sites. At present, data supporting the monitoring function comes from U.S. military assets located throughout the globe. In the future, sources for this data could also include non-military and non-U.S. monitoring sites. If this is to occur, means of insuring data authentication must be established.

1.3 DOCUMENT DESCRIPTION

The purpose of civil monitoring is to ensure that civil GPS performance standards are achieved, to aid satellite operations in minimizing adverse impacts to users, and to assess the level of performance of the GPS SPS. Civil monitoring is *not* intended to provide application specific monitoring such as those employed in providing safety of life integrity monitoring services.

In support of its objectives, this CMPS provides a set of metrics for measuring GPS performance relative to standards defined in U.S. Government policy and high-level system definitions. These policy statements and definitions include the Federal Radionavigation Plan, the Standard Positioning Service Performance Standard, the Wide Area Augmentation System (WAAS) Performance Standard (WAAS PS), and the Navstar GPS Space Segment/Navigation User Interfaces (IS-GPS-200) for the definition of currently available signals and services. Also included are the Navstar GPS Space Segment/User Segment L5 Interfaces (IS-GPS-705) and the Navstar GPS Space Segment/User Segment L1C Interfaces (IS-GPS-800) for the definition of future signals and services. To the extent practicable, each defined metric is traceable to one or more specifications of performance. In cases where the metric is an indirect measure of the performance, the connection between the metric and the standard is explained and the threshold and/or goal necessary to achieve acceptable performance provided.

This CMPS also defines the scope and range of monitoring needs not otherwise documented but reasonably expected by civil users. These include the ability of the service to detect defects in signal and data, the rapid report of anomalous service behavior to satellite operations for resolution, and notification to users of the causes and effects of such anomalies for their various service types (e.g., positioning, timing, and navigation).

It is important to note that the SPS PS defines only one civil service associated with GPS, the service delivered via the L1 C/A code. The word "service" should be interpreted to mean the L1 C/A SPS. At the same time, it is recognized that there are a variety of means for using GPS beyond the official definition of SPS. While it is beyond the scope of this document to define new classes of service, it is possible to define monitoring criteria on a signal-by-signal basis. In this document, it is assumed that users implementing approaches beyond basic GPS have (or will) base their GPS implementations on the signal specifications contained in the referenced documents (specifically the IS-GPS-200, IS-GPS-705, and IS-GPS-800). This CMPS then defines signal monitoring requirements for L1 C/A, L1C, L2C, and L5 that assure the signal specifications are met based on the signal descriptions in these interface documents. In this same spirit, this CMPS does not address augmentation services; however there are several classes of civil users that depend on semi-codeless receiver technologies. The WAAS PS is included as a reference because it states explicit assumptions about the GPS signal that are relevant to semi-codeless receiver technologies. To maintain a point-of-view consistent with both the SPS PS and the interface specifications, the performance metrics defined in this CMPS are defined in terms of the GPS Signal In Space (SIS) without considering the impact of atmospheric propagation (e.g. ionospheric and tropospheric errors) or local reception (e.g. terrain masking or interference).

Some performance metrics are known to be of interest, but are not included in the current draft of this document in order to remain within the definitions currently in existing policy and system documentation. It is anticipated that metrics associated with items such as these will be included in future revisions of this CMPS as policy evolves and reference documents listed in Section 2 are updated.

The metrics defined in this CMPS are of immediate use to the GPS service provider (the U.S. Air Force) and to USG agencies such as the U.S. Coast Guard Navigation Center and the FAA National Operations Control Center that have responsibility for communicating with GPS end users. All users will benefit from: (1) reduced outage times by timely notification to the GPS operations when anomalies occur; and (2) long-term assurance that U.S. Government commitments regarding GPS service levels are consistently met.

In addition, this CMPS addresses the need to assess the level of performance of the civil services, even when the services exceed the minimum standards. As a result, this CMPS contains requirements for archives of key data and events to support future improvements in GPS service and to respond to external queries about actual GPS service levels.

The requirements presented in Section 3 represent the monitoring requirements related to the SPS and the GPS signals used by the civil community. While these requirements are of interest to the civil community, many of them are of equal interest to the military. For example: (1) the requirements associated with constellation management are applicable to both groups; and (2) the requirements associated with L1 C/A are of interest to the military due to the large number of legacy user equipment receivers that require L1 C/A in order to acquire the PPS. Therefore, there is a significant amount of overlap between the civil monitoring requirements and those of general interest. This is illustrated in Figure 1-1.

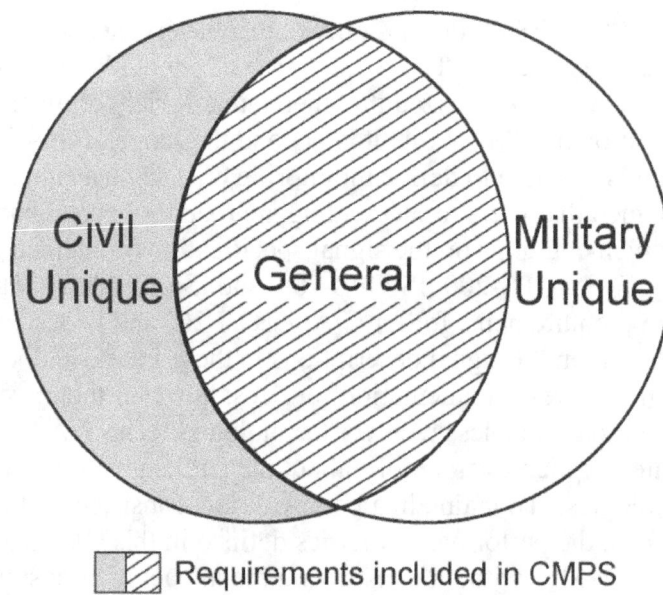

Figure 1-1 – Partitioning of Monitoring Requirements

This partitioning of requirements may have practical implications. In the current "U.S. Space-Based Positioning, Navigation, and Timing Policy", it is noted that "...*civil signal performance monitoring...will be funded by the agency or agencies requiring those services or capabilities, including out-year procurement and operations costs.*" Therefore, funding for the development and operation of monitoring capabilities may fall under the jurisdiction of different organizations depending on whether the requirement is related to a particular user community or is perceived to be of general importance.

To illustrate how the requirements from Section 3 may be partitioned, Section 4 contains a table that partitions the requirements into civil-unique and general categories.

This document does not state how civil monitoring will be implemented nor will it address the monitoring system architecture. In particular, no assumption is made regarding the level of automation of the service. The purpose of this document is to provide a comprehensive statement of civil service and signal monitoring requirements. Many of the requirements may be incorporated into the next generation operational control system (OCX) while other requirements may be allocated to other government entities for implementation.

2 APPLICABLE DOCUMENTS

2.1 GENERAL
This section lists source documents for the requirements delineated in Sections 3 and 4.

2.2 GOVERNMENT DOCUMENTS

2.2.1 Specifications, standards, and handbooks

IS-GPS-200D, IRN 001 7 March 2006	Navstar GPS Space Segment/Navigation User Interfaces
ICD-GPS-240 1 October 2004	Navstar GPS Control Segment to User Support Community Interfaces
IS-GPS-705, IRN 001, 002, 003 22 September 2005	Navstar GPS Space Segment/ User Segment L5 Interfaces
IS-GPS-800 4 September 2008	Navstar GPS Space Segment/ User Segment L1C Interfaces
SS-SYS-800C 14 Aug 2008	GPS III System Specification for the Global Positioning System Wing (GPSW) (FOUO)
September 2008	Global Positioning System Standard Positioning Service Performance Standard – 4th Edition
October 2008	Wide Area Augmentation Service Performance Standard – 1st Edition

2.2.2 Other Government documents, drawings, and publications

Jan 2009	2008 Federal Radionavigation Plan DOT-VNTSC-RITA-08-02/ DOD-4650.5
3 November 2008	Memorandum of Agreement, between the Department of Defense and the Department of Transportation, Civil Use of the Global Positioning System
January 2003	GPS Integrity Failure Modes and Effects Analysis (IFMEA) 2002 Final Report (performed by Volpe Center for the Interagency GPS Executive Board)
26 July 2004	IFOR Proposed New Operational Requirement, Daniel P.

GPS Civil Monitoring Performance Specification

	Salvano, FAA
26 July 2004	Attachment 1 to April 25, 2003 IFOR Proposed New Operational Requirement, Aviation Backward Compatibility, Detailed Requirements, Daniel P. Salvano, FAA
23 September 2008	"Preservation of Continuity for Semi-Codeless GPS Applications", Federal Register

2.3 NON-GOVERNMENT DOCUMENTS

May 2008	Recommendations on Digital Distortion Requirements for the Civil GPS Signals, IEEE/ION PLANS 2008, May 6-8, 2008, Monterey, CA
20 November 2008	ICAO SARPs, Annex 10, Attachment D. Information and Material for Guidance in the Application of the GNSS Standards and Recommended Practices (SARPs)

3 REQUIREMENTS

Except where explicitly noted, civil monitoring shall meet the requirements stipulated herein. The origin of each requirement is identified inside square brackets following the statement of the requirement. For the purposes of identifying the origin of the requirements, the following abbreviations are used

- SPS PS – Standard Positioning Service Performance Standard
- WAAS PS – WAAS Performance Standard
- SS-SYS-800 - GPS III System Specification
- IS-GPS-200 – Navstar GPS Space Segment/Navigation User Interfaces
- ICD-GPS-240 - Navstar GPS Control Segment to User Support Community Interfaces
- IS-GPS-705 – Navstar GPS Space Segment/ User Segment L5 Interfaces
- IS-GPS-800 – Navstar GPS Space Segment/User Segment L1C Interfaces
- IFOR Prop - IFOR Proposed New Operational Requirement, July 26, 2004
- SARPs-10D – ICAO SARPs, Annex 10, Attachment D

System attributes used in defining the requirements in this section are based on the definitions found in Section 5.5.

3.1 SYSTEM PERFORMANCE MONITORING REQUIREMENTS

All monitoring requirements defined in Sections 3.1.1 through 3.1.8 refer to the SPS as defined in the September 2008 SPS PS.

3.1.1 Verification of Constellation Management Standards

Civil monitoring shall verify that:

a. the argument of latitude of each SV is maintained to within +/- 4° of the nominal value [SPS PS Section 3.2], and
b. the eccentricity of each satellite is within the required tolerance of 0.00-0.03 [SPS PS Section 3.2].

3.1.2 Verification of Signal in Space Coverage Standards

The following monitoring requirements apply to coverage standards. Civil monitoring shall verify that:

a. the terrestrial service volume coverage per satellite is 100% [SPS PS Section 3.3.1], and
b. the terrestrial service volume constellation coverage is 100% [SPS PS Section 3.3.2].

3.1.3 Verification of Signal in Space Accuracy Standards

The following monitoring requirements apply to 'healthy' satellites as defined in the SPS PS. Requirements a.-g. are specific to the SPS SIS. Civil monitoring shall verify that:

a. the 95th % global average SPS SIS User Range Error (URE) for each SV during Normal Operations over all ages of data (AOD) is less than or equal to 7.8 meters [SPS PS Section 3.4.1],
b. the 95th % global average SPS SIS URE for each SV during Normal Operations at zero AOD is less than or equal to 6 meters [SPS PS Section 3.4.1],
c. the 95th % global average SPS SIS URE for each SV during Normal Operations at any AOD is less than or equal to 12.8 meters [SPS PS Section 3.4.1],
d. the percentage of time the SPS SIS URE is 30 meters or less with daily percentage values averaged over a year is greater than or equal to 99.94% [SPS PS Section 3.4.1],
e. the percentage of time the SPS SIS URE is 30 meters or less for the worst-case point within the service volume with daily percentage values averaged over a year is greater than or equal to 99.79% [SPS PS Section 3.4.1],
f. the 95th % global average SPS SIS User Range Rate Error (URRE) over any 3-second interval during Normal Operation at any AOD is less than or equal to 0.006 m/sec [SPS PS Section 3.4.2],
g. the 95th % global average SPS SIS User Range Acceleration Error (URAE) over any 3-second interval during Normal Operation at any AOD is less than or equal to 0.002 m/sec/sec 95% [SPS PS Section 3.4.3], and
h. the 95th % global average Coordinated Universal Time Offset Error (UTCOE) during Normal Operation at any AOD is less than or equal to 40 ns [SPS PS Section 3.4.4].

3.1.4 Verification of Signal in Space Reliability Standards

The following monitoring requirements apply to 'healthy' satellites as defined in the SPS PS. Civil monitoring shall verify that:

a. the percentage of time the SIS Instantaneous URE of the GPS III civil signals (L1 C/A, L2C, L5, L1C) exceeds the not-to-exceed (NTE) tolerance without a timely alert is less than or equal to 1×10^{-5} per hour per SV [SPS PS Section 3.5.1][1],
b. the percentage of time the percentage of time the SIS Instantaneous UTCOE of the GPS III civil signals (L1 C/A, L2C, L5, L1C) exceeds the not-to-exceed (NTE) tolerance without a timely alert is less than or equal to 1×10^{-5} per hour per SV [SPS PS Section 3.5.4],
c. the percentage of time the SIS User Range Rate Error (URRE) of the GPS III civil signals (L1 C/A L2C, L5, L1C) exceeds 2.0 cm per second is less than or equal to 1×10^{-5}/sample where the sample size is defined as any 3 second interval. [SS-SYS-800, 3.2.1.4], and
d. the percentage of time the SIS User Range Acceleration Error (URAE) of the GPS III civil signals (L1 C/A L2C, L5, L1C) exceeds 7 mm per second^2 is less than or equal to 1×10^{-5}/sample where the sample size is defined as any 3 second interval. [SS-SYS-800, 3.2.1.5].

3.1.5 Verification of Signal in Space Continuity Standards

The external notices described below and their verification are the responsibility of the SPS provider. Civil monitoring shall:

---

[1] The SPS PS only defines this for Single-Frequency C/A-Code. It is anticipated the remaining civil codes will eventually be addressed by a similar, or more stringent, specification.

a. verify that notice is issued no less than 48 hours in advance of any planned disruption of the SPS (defined to be periods in which the GPS is not capable of providing SPS as specified in the SPS Performance Standard) as specified in the SPS PS [SPS PS Section 3.6.3],
b. verify that notice is issued no less than 48 hours in advance of scheduled change in constellation operational status that affects the service being provided to GPS users [SPS PS Section 3.6.3],
c. monitor the time to issue a notification of unscheduled outages or problems [SPS PS Section 3.6.3], and
d. verify that the fraction of hours in a year that the SPS SIS from any slot is lost due to unscheduled failure is less than 0.0002 [SPS PS Section 3.6.1].

3.1.6 Verification of Signal in Space Availability Standards

Civil monitoring shall verify that:

a. the fraction of time over a one year period that a slot in the baseline 24-Slot configuration is occupied by a SV broadcasting a Healthy SPS SIS is greater than or equal to 0.957 [SPS PS Section 3.7.1][2],
b. the fraction of time over a one year period that 21 satellites in the baseline 24 –Slot configuration and set healthy and broadcasting a navigation signal is greater than or equal to 0.98 [SPS PS Section 3.7.2],
c. the fraction of time over a one year period that at least 20 slots of the baseline 24 –Slot configuration will be occupied by a SV or SVs broadcasting a Healthy SPS SIS is greater than or equal to 0.99999 [SPS PS Section 3.7.2], and
d. the fraction of time over a 24-hour period that 24 operational satellites are available on orbit is greater than or equal to 0.95 [SPS PS Section 3.7.3].

3.1.7 Verification of Position/Time Domain Availability

Civil monitoring shall verify that:

a. the percentage of time the constellation's global Position Dilution of Precision (PDOP) value is 6 or less is greater than or equal to 98% within the service volume over any 24-hour interval [SPS PS Section 3.8.1][3],
b. the percentage of time the constellation's worst site PDOP value is 6 or less is greater than or equal to 88% within the service volume over any 24-hour interval [SPS PS Section 3.8.1][3],
c. availability 95% horizontal accuracy of 17 meters is greater than or equal to 99% in any 24-hour interval for an average location within the service volume considering only the SIS component of accuracy [SPS PS Section 3.8.2],
d. availability 95% vertical accuracy of 37 meters is greater than or equal to 99% in any 24-hour interval for an average location within the service volume considering only the SIS component of accuracy [SPS PS Section 3.8.2],

---

[2] Throughout this section, a "slot in the baseline 24-Slot configuration" means either one SV in the nominal slot location, or two SVs occupying the nominal locations in the expanded slot.
[3] See Section 5.4.6 for the process for DOP assessment.

e. availability 95% horizontal accuracy of 17 meters is greater than or equal to 90% in any 24-hour interval at the worst-case location in the service volume considering only the SIS component of accuracy [SPS PS Section 3.8.2], and
f. availability 95% vertical accuracy of 37 meters is greater than or equal to 90% in any 24-hour interval at the worst-case location in the service volume considering only the SIS component of accuracy [SPS PS Section 3.8.2].

3.1.8 Verification of Position/Time Domain Accuracy

Civil monitoring shall verify that:

a. the global average horizontal positioning domain accuracy measured over a 24-hour interval is less than or equal to 9 meters 95% considering only signal in space errors and using an all-in-view receiver algorithm [SPS PS Section 3.8.3][4],
b. the global average vertical positioning domain accuracy measured over a 24-hour interval is less than or equal to 15 meters 95% considering only signal in space errors and using an all-in-view receiver algorithm [SPS PS Section 3.8.3][4],
c. the horizontal positioning domain accuracy for the worst site in the service volume measured over a 24-hour interval is less than or equal to 17 meters 95% considering only signal in space errors and using an all-in-view receiver algorithm [SPS PS Section 3.8.3][4],
d. the vertical positioning domain accuracy for the worst site in the service volume measured over a 24-hour interval is less than or equal to 37 meters 95% considering only signal in space errors and using an all-in-view receiver algorithm [SPS PS Section 3.8.3][4], and
e. the time transfer accuracy is less than or equal to 40 ns 95%, averaged over the service volume over any 24-hour period assuming an all-in-view receiver at a surveyed location and considering only the SIS component of accuracy [SPS PS Section 3.8.3][4].

3.2 CIVIL SIGNAL MONITORING REQUIREMENTS

3.2.1 Verification of Civil Ranging Codes

Civil signal monitoring shall:

a. detect and monitor instances of non-standard code transmission of the L1 C/A code [IS-GPS-200 Section 3.2.1],
b. detect and monitor instances of non-standard code transmission of the L2 civil-moderate (CM) and L2 civil-long (CL) code [IS-GPS-200 Section 3.2.1],
c. track the C/A code on L1 regardless of status of the health bits in the navigation message[5] [IS-GPS-200 Section 3.2.1.3],
d. track the CM-code on L2 regardless of status of the health bits in the navigation message[5] [IS-GPS-200 Section 3.2.1.4],

---

[4] See Section 5.4.7 for definitions related to averaging and accuracy assessment.
[5] The term "track" means to acquire the signal, collect measurements, and collect the navigation message data (if available). In these cases, the IS reference in the square bracket denotes the section of the IS where the relevant signal is described and is not an indication of traceability.

e. track the CL-code on L2 regardless of status of the health bits in the navigation message[5] [IS-GPS-200 Section 3.2.1.5],
f. detect when any satellite transmits pseudorandom noise (PRN) codes 33 through 37 [IS-GPS-200 Table 3-1],
g. detect and monitor instances of non-standard code transmission of the L5 I5 code [IS-GPS-705 Section 3.2.1.2],
h. detect and monitor instances of non-standard code transmission of the L5 Q5 code [IS-GPS-705 Section 3.2.1.2],
i. track the I5-code on L5 regardless of status of the health bits in the navigation message[5] [IS-GPS-705 Section 3.2.1],
j. track the Q5-code on L5 regardless of status of the health bits in the navigation message[5] [IS-GPS-705 Section 3.2.1],
k. detect and monitor instances of non-standard code transmission of the L1CP code [IS-GPS-800 Section 3.2.2.2],
l. detect and monitor instances of non-standard code transmission of the L1CD code [IS-GPS-800 Section 3.2.2.2],
m. track the L1C code regardless of status of the health bits in the navigation message[5] [IS-GPS-800 Section 3.2.2.1],
n. detect and monitor instances of carrier phase tracking discontinuities for L1, L2, and L5 [see 5.4.5],
o. detect when the L1 C/A navigation message is not synchronized with the L1 C/A code [IS-GPS-200 Section 3.3.4, Fig 3-16],
p. detect when the L2C navigation message is not synchronized with the L2C code [IS-GPS-200 Section 3.3.3.1.1],
q. detect when the L5 navigation message is not synchronized with the L5 code [IS-GPS-705 Section 3.3.3.1.1], and
r. detect when the L1C navigation message is not synchronized with the $L1C_D$ code [IS-GPS-800 Section 3.5.2].

3.2.2 Civil Signal Quality Monitoring

Civil signal monitoring shall:

a. verify the received minimum radio frequency (RF) signal strength on L1 C/A is at or above -158.5 dBW for each space vehicle (SV) transmitting a healthy L1 C/A signal (see Section 5.4.1) [IS-GPS-200 Section 3.3.1.6][6],
b. verify the received minimum RF signal strength on L2C is at or above -160.0 dBW for each SV transmitting a healthy L2C signal (see Section 5.4.1) [IS-GPS-200 Section 3.3.1.6][6],
c. verify the received minimum RF signal strength on L5/I5 is at or above -157.9 dBW for each GPS IIF SV transmitting a healthy L5 signal (see Section 5.4.1) [IS-GPS-705 Section 3.3.1.6][6],

---

[6] See Section 5.4.1 and use case in Section 5.2.7. The term "healthy" is to be interpreted as defined in the SPS PS (Section 2.3.2). The referenced IS paragraphs contain additional constraints that must be considered (e.g. elevation angle, assumptions regarding the antenna).

GPS Civil Monitoring Performance Specification

d. verify the received minimum RF signal strength on L5/I5 from GPS III SVs is at or above -157 dBW for each GPS III SV transmitting a healthy L5 signal (see Section 5.4.1) [DRAFT IS-GPS-705 Section 3.3.1.6][6],
e. verify the received minimum RF signal strength on L5/Q5 is at or above -157.9 dBW for each GPS IIF SV transmitting a healthy L5 signal (see Section 5.4.1) [IS-GPS-705 Section 3.3.1.6][6],
f. verify the received minimum RF signal strength on L5/Q5 from GPS III SVs is at or above -157 dBW for each GPS III SV transmitting a healthy L5 signal (see Section 5.4.1) [DRAFT IS-GPS-705 Section 3.3.1.6][6],
g. verify the terrestrial received minimum RF signal strength on L1C is at or above -157.0 dBW for each SV transmitting a healthy L1C signal (see Section 5.4.1) [IS-GPS-800 Section 3.2.1.9][6],
h. verify the orbital received minimum RF signal strength on L1C is at or above -182.5 dBW for each SV transmitting a healthy L1C signal (see Section 5.4.1) [IS-GPS-800 Section 3.2.1.9][6],
i. continuously monitor the C/N_0 from L1 C/A for each SV transmitting a healthy signal and report significant drops (see Section 5.4.2) [IS-GPS-200 Section 3.3.1.6][7],
j. continuously monitor the received C/N_0 from L2C for each SV transmitting a healthy L2C signal and report significant drops (see Section 5.4.2) [IS-GPS-200 Section 3.3.1.6][7],
k. continuously monitor the received C/N_0 from L5 for each SV transmitting a healthy L5 signal and report significant drops (see Section 5.4.2) [IS-GPS-705 Section 3.3.1.6][7],
l. continuously monitor the received C/N_0 from L1C for each SV transmitting a healthy L1C signal and report significant drops (see Section 5.4.2) [IS-GPS-800 Section 3.2.1.9][7],
m. verify code-carrier divergence in the L1 C/A signal is less than 6.1 meters over any period of time T between 100 seconds and 7200 seconds (see Section 5.4.3) [IFOR Prop 13a],
n. verify code-carrier divergence in the L2C signal is less than 6.1 meters over any period of time T between 100 seconds and 7200 seconds (see Section 5.4.3) [IFOR Prop 13a],
o. verify code-carrier divergence in the L5 signal is less than 6.1 meters over any period of time T between 100 seconds and 7200 seconds (see Section 5.4.3.) [IFOR Prop 13a],
p. verify code-carrier divergence in the L1C signal is less than 6.1 meters over any period of time T between 100 seconds and 7200 seconds (see Section 5.4.3) [IFOR Prop 13a],
q. verify the average time difference between the L1 C/A code and L1 P(Y) code transitions does not exceed 10 ns (two sigma) [Section 3.3.1.8],
r. verify the mean group differential delay between the L1 P(Y) and L2C codes does not exceed 15 ns [IS-GPS-200 Section 3.3.1.7.2],
s. verify the mean group differential delay between the L1 P(Y) and L5 codes does not exceed 30 ns [IS-GPS-705 Section 3.3.1.7.2],
t. verify the mean group differential delay between the L1 P(Y) and L1C codes does not exceed 15 ns [IS-GPS-800 Section 3.2.1.8.2],
u. verify stable 90 degree phase offset (+/- 100 milliradians) between L1 C/A and L1 P(Y) code carriers with C/A lagging P(Y) [IS-GPS-200 Section 3.2.3 and 3.3.1.5, WAAS PS Appendix A.3.2],
v. verify stable 90 degree phase offset (+/- 100 milliradians) between L2C and L2 P(Y) code carriers with L2C lagging L2 P(Y) [IS-GPS-200 Section 3.2.3, WAAS PS Appendix A.3.2],

---

[7] See Section 5.4.2 and use case in Section 5.2.7.

w. verify the magnitude of an L1 C/A code chip's lead and lag variation from a square wave does not exceed 0.12 chips [SARPs-10D 8.4],
x. verify the magnitude of an L2 C code chip's lead and lag variation from a square wave does not exceed 0.02 chips [see Section 5.4.4],
y. verify the magnitude of an L5I and L5Q code chip's lead and lag variation from a square wave does not exceed 0.02 chips [see Section 5.4.4],
z. verify the magnitude of an L1C code chip's lead and lag variation from a square wave does not exceed 0.05 chips [see Section 5.4.4], and
aa. Detect and monitor instances when the transient (unit step) response for each bit transition exceeds the limits defined in SARPS Threat Model B [SARPs-10D 8.5.2].

3.2.3 Verification of Signal Characteristics for Semi-Codeless Tracking

Civil signal monitoring shall:

a. verify L2 modulated with the same P(Y) as L1 [WAAS PS, Appendix A.3.2],
b. verify the same navigation data is broadcast on both L1 P(Y) and L2 P(Y) [WAAS PS, Appendix A.3.2],
c. verify L1-L2 differential bias stability less than 3 ns, 2 sigma over any 5 minute interval [IS-GPS-200, Section 3.3.1.7.2], and
d. verify that the group delay differential between the L1 P(Y) and L2 P(Y) does not exceed 15.0 nanoseconds [IS-GPS-200, Section 3.1.7.2, WAAS PS A.3.2].

3.2.4 Verification of Navigation Message

The following requirements address the correctness of the navigation message data with respect to the definitions contained in the relevant interface documents. Where the term "correctly set" is used, it may be interpreted to mean either "consistent with the intent of the interface definition" and "within an acceptable set or range of values as established for the monitoring function, based on the nature of each parameter (or parameter set) and the manner in which it is determined". There is no intent to imply that civil signal monitoring must perform an independent solution for each SV and compare the results to the broadcast navigation message.

3.2.4.1 Verification of L1 C/A Navigation Message

Civil signal monitoring shall:

a. verify the GPS time scale is within one microsecond of the Coordinated Universal Time (UTC) when adjusted by the leap Second Correction [IS-GPS-200 Section 3.3.4],
b. detect the transmission of alternating ones and zeroes in words 3 through 10 in place of normal L1 C/A navigation message (NAV) data [IS-GPS-200 Section 20.3.2],
c. verify the correct time of week count is present in the handover word (HOW) [IS-GPS-200 Section 20.3.3.2],
d. verify the GPS week number is incremented at each end/start of week epoch [IS-GPS-200 Section 20.3.3.3.1.1],
e. verify that the clock parameters in subframe 1 are correctly set [IS-GPS-200 Section 20.3.3.3.1],

f. verify that the ephemeris parameters in subframes 2 and 3 is correctly set [IS-GPS-200 Section 20.3.3.4.1],
g. verify that the time of ephemeris (t_{oe}) value, for at least the first data set transmitted by an SV after an upload, is different from that transmitted prior to the cutover as specified in IS-GPS-200 [IS-GPS-200 Section 20.3.3.4.1, 20.3.4.5],
h. verify that each SV is in "normal operations" mode by verifying the fit interval flag is set to zero (0) and the value of the 8 least significant bits of the IODC are in the range 0-239 [IS-GPS-200 Section 20.3.3.4.3.1, 20.3.4.4],
i. verify that the almanac message for any dummy SVs contains alternating ones and zeros with valid parity [IS-GPS-200 Section 20.3.3.5.1.2],
j. verify that the almanac parameters are updated by the Control Segment at least once every 6 days as specified in IS-GPS-200 [IS-GPS-200 Section 20.3.3.5.1.2],
k. verify that the almanac reference week and time of almanac (t_{oa}) define a time that is between the time of transmission and a time no more than 3.5 days in the future from the time of transmission [IS-GPS-200 Section 20.3.3.5.2.2],
l. verify that UTC parameters are correctly set as specified in IS-GPS-200 [IS-GPS-200 Section 20.3.3.5.2.4],
m. verify that the absolute value of the difference between the untruncated week number (WN) and truncated leap second week number (WNt) values does not exceed 127 when Δt_{LS} and Δt_{LSF} differ [IS-GPS-200 Section 20.3.3.5.2.4],
n. verify that the reference time for UTC is correctly set [IS-GPS-200 Section 20.3.3.5.2.4],
o. verify that the UTC parameters provided via L1 C/A navigation message are updated by the Control Segment at least once every 6 days [IS-GPS-200 Section 20.3.3.5.1.6],
p. verify that the single frequency ionospheric parameters are correctly set [IS-GPS-200 Section 20.3.3.5.1.7],
q. verify that the single frequency ionospheric data are updated by the Control Segment at least once every 6 days [IS-GPS-200 Section 20.3.3.5.1.7],
r. verify that the almanac is correctly set [IS-GPS-200 Section 20.3.3.5.2.1],
s. verify that the almanac time parameters provided a statistical URE component that is less than 135 m one sigma [IS-GPS-200 Section 20.3.3.5.2.3],
t. verify that the transmitted issue of data clock (IODC) is different from any value transmitted by the SV during the time period specified in IS-GPS-200 [IS-GPS-200 Section 20.3.4.4],
u. verify that the transmitted issue of data ephemeris (IODE) is different from any value transmitted by the SV during the time period specified in IS-GPS-200 [IS-GPS-200 Section 20.3.4.4],
v. verify that the transmitted IODC values obey the assignment rules specified in IS-GPS-200 (assuming normal operations are in effect) [IS-GPS-200 Section 20.3.4.4, Table 20-XI, 20-XII],
w. verify that the group delay differential terms are correctly set [IS-GPS-200 Section 3.3.1.7.2],
x. verify that the week number of transmission in the L1 C/A navigation message is correctly set [IS-GPS-200 Section 20.3.3.3.1.1], and
y. verify that carrier phase discontinuities do not result in unintended navigation bit inversion [IS-GPS-200 Section 3.2.2].

3.2.4.2 Verification of L1C Navigation Message

Civil signal monitoring shall:

a. detect the transmission of alternating ones and zeros in the L1C subframe 2 and/or subframe 3 navigation message [IS-GPS-800 Section 3.5.1],
b. verify a correct Time of Interval (TOI) count is present in each subframe 1 [IS-GPS-800, 3.2.3.1, 3.5.1],
c. verify the GPS week number is incremented at each end/start of week epoch [IS-GPS-800 Section 3.5.3.1],
d. verify that the ephemeris and clock parameters in subframe 2 are correctly set [IS-GPS-800 Section 3.5.3.6],
e. verify that the time of ephemeris (t_{oe}) value, for at least the first data set transmitted by an SV after an upload, is different from that transmitted prior to the cutover as specified in IS-GPS-800 [IS-GPS-800 Section 3.5.3],
f. verify that the almanac reference week and time of almanac (t_{oa}) define a time that is between the time of transmission and a time no more than 3.5 days in the future from the time of transmission [IS-GPS-800 Section 3.5.4.3.2],
g. verify that UTC parameters are correctly set as specified in IS-GPS-800 [IS-GPS-800 Section 3.5.4.1.1],
h. verify that the UTC parameters are updated by the Control Segment at least once every 3 days [IS-GPS-800 Section 3.5.4.1.1],
i. verify that the single frequency ionospheric parameters are correctly set [IS-GPS-800 Section 3.5.4.1.2],
j. verify that the single frequency ionospheric data are updated by the Control Segment at least once every 6 days [IS-GPS-800 Section 3.5.4.1.2],
k. verify that the almanac is correctly set [IS-GPS-800 Section 3.5.4.3],
l. verify that the reference time for UTC is correctly set [IS-GPS-800 Section 3.5.4.1.1.1],
m. verify that the group delay differential terms are set correctly [IS-GPS-800 Section 3.2.1.8.1, 3.2.1.8.2, 3.5.3.9],
n. verify that the reduced almanac parameters are updated by the Control Segment at least once every 3 days [IS-GPS-800 Section 3.5.4.3.5],
o. verify that the almanac reference week is correctly set [IS-GPS-800 Section 3.5.4.3.1],
p. verify the GPS time scale is within 50 ns of the Coordinated Universal Time (UTC) when adjusted by the leap Second Correction (when GPS III is operational) [IS-GPS-800 Section 3.4.1],
q. verify that any change in the subframe 2 ephemeris and clock data occurs in conjunction with a change in the time of ephemeris (t_{oe}) value [IS-GPS-800 Section 3.5.3],
r. verify L1C subframe 3, page 1 (UTC and IONO) is updated on each SV at least once every three days [IS-GPS-800 Section 3.5.4.1.1],
s. verify that the GGTO parameters are correctly set [IS-GPS-800 Section 3.5.4.2.1],
t. verify that the EOP parameters are correctly set [IS-GPS-800 Section 3.5.4.2.2],
u. verify that the differential correction parameters are correctly set [IS-GPS-800 Section 3.5.4.4.1],

v. verify that the week number of transmission in the L1C navigation message is correctly set [IS-GPS-800 Section 3.5.3.1], and
w. verify that carrier phase discontinuities do not result in unintended navigation bit inversion [IS-GPS-800 Section 3.3].

3.2.4.3 Verification of L2C Navigation Message

Civil signal monitoring shall:

a. verify the correct TOW is present in each message [IS-GPS-200 30.3.3],
b. detect default message data [IS-GPS-200 Section 30.3.3],
c. verify the GPS week number is incremented at each end/start of week epoch [IS-GPS-200 Section 30.3.3.1.1.1],
d. verify that the ephemeris parameters in Message Types 10 and 11 are correctly set [IS-GPS-200 Section 30.3.3.1.1],
e. verify that the clock parameters in Message Types 30-37 are correctly set [IS-GPS-200 Section 30.3.3.2.1],
f. verify that the time of ephemeris (t_{oe}) value, for at least the first data set transmitted by an SV after an upload, is different from that transmitted prior to the cutover as specified in IS-GPS-200 [IS-GPS-200 Section 30.3.3.1.1],
g. verify that the almanac parameters are updated by the Control Segment at least once every 3 days as specified in IS-GPS-200 [IS-GPS-200 Section 30.3.3.4.6.1],
h. verify that the almanac reference week and time of almanac (t_{oa}) define a time that is between the time of transmission and a time no more than 3.5 days in the future from the time of transmission [IS-GPS-200 Section 30.3.3.4.1],
i. verify that UTC parameters are correctly set as specified in IS-GPS-200 [IS-GPS-200 Section 30.3.3.6.1],
j. verify that the UTC parameters are updated by the Control Segment at least once every 3 days [IS-GPS-200 Section 30.3.3.6.1],
k. verify that the single frequency ionospheric parameters are correctly set [IS-GPS-200 Section 30.3.3.3.1],
l. verify that the single frequency ionospheric data are updated by the Control Segment at least once every 6 days [IS-GPS-200 Section 30.3.3.3.1.2],
m. verify that the almanac is correctly set [IS-GPS-200 Section 30.3.3.4],
n. verify that the absolute value of the difference between the untruncated week number (WN) and truncated leap second week number (WNt) values does not exceed 127 when Δt_{LS} and Δt_{LSF} differ [IS-GPS-200 Section 20.3.3.5.2.4],
o. verify that the reference time for UTC is correctly set [IS-GPS-200 Section 20.3.3.5.2.4],
p. verify that the group delay differential terms are set correctly [IS-GPS-200 Section 30.3.3.3.1.1],
q. verify that the reduced almanac parameters in Message Types 12, 31, and 37 are updated by the Control Segment at least once every 3 days [IS-GPS-200 Section 30.3.3.4.6.1],
r. verify that the week number of transmission in the L2C navigation message is correctly set [IS-GPS-200 Section 30.3.3.1.1.1], and
s. verify that carrier phase discontinuities do not result in unintended navigation bit inversion [IS-GPS-200 Section 3.2.2].

3.2.4.4 Verification of L5 Navigation Message

Civil signal monitoring shall:

a. verify the correct TOW count is present in each message [IS-GPS-705 20.3.3],
b. detect default message data [IS-GPS-705 Section 20.3.2, 20.3.3],
c. verify the GPS week number is incremented at each end/start of week epoch [IS-GPS-705 Section 20.3.3.1.1.1],
d. verify that the ephemeris parameters in Message Types 10 and 11 are correctly set [IS-GPS-705 Section 20.3.3.1.1],
e. verify that the clock parameters in Message Types 30-37 are correctly set [IS-GPS-705 Section 20.3.3.2.1],
f. verify that the time of ephemeris (t_{oe}) value, for at least the first data set transmitted by an SV after an upload, is different from that transmitted prior to the cutover as specified in IS-GPS-705 [IS-GPS-705 Section 20.3.3.1],
g. verify that the almanac parameters in the navigation message are updated by the Control Segment at least once every 3 days as specified in IS-GPS-705 [IS-GPS-705 Section 20.3.3.4.6.1],
h. verify that the almanac reference week and time of almanac (t_{oa}) define a time that is between the time of transmission and a time no more than 3.5 days in the future from the time of transmission [IS-GPS-705 Section 20.3.3.4.1],
i. verify that UTC parameters are correctly set as specified in IS-GPS-705 [IS-GPS-705 Section 20.3.3.6.1],
j. verify that the UTC parameters are updated by the Control Segment at least once every 3 days [IS-GPS-705 Section 20.3.3.6.1],
k. verify that the single frequency ionospheric parameters are correctly set [IS-GPS-705 Section 20.3.3.3.1],
l. verify that the single frequency ionospheric data are updated by the Control Segment at least once every 6 days [IS-GPS-705, Section 20.3.3.3.1.3],
m. verify that the almanac is correctly set [IS-GPS-705 Section 20.3.3.4.5, 20.3.3.4.6],
n. verify that the reference time for UTC is correctly set [IS-GPS-705 Section 20.3.3.6.2],
o. verify that the L1-L2 group delay differential terms are set correctly [IS-GPS-705 Section 20.3.3.3.1.1],
p. verify that the L1-L5 group delay differential terms in L5 messages are set correctly [IS-GPS-705 Section 3.3.1.7.1, 3.3.1.7.2, 20.3.3.3.1.2],
q. verify that the almanac reference week is correctly set [IS-GPS-705 Section 20.3.3.4.1].
r. verify that the week number of transmission in the L5 navigation message is correctly set [IS-GPS-705 Section 20.3.3.1.1.1],
s. verify the GPS time scale is within 90 ns of the Coordinated Universal Time (UTC) when adjusted by the leap Second Correction [IS-GPS-705 Section 3.3.4], and
t. verify that carrier phase discontinuities do not result in unintended navigation bit inversion [IS-GPS-705 Section 3.2.2].

3.3 VERIFICATION OF GPS III INTEGRITY

GPS III Integrity requirements are still evolving. Eventually these may imply monitoring requirements. There are no additional monitoring requirements associated with GPS III integrity at this time.

3.4 NON-BROADCAST DATA MONITORING REQUIREMENTS

Civil monitoring shall:

a. verify each Notice Advisory to Navstar User (NANU) message created by the GPS service provider meets the format and transmission requirements specified in ICD-GPS-240 [ICD-GPS-240 Section 3.2.2.1],
b. verify each Operational Advisory message created by the GPS service provider meets the format and transmission requirements specified in ICD-GPS-240 [ICD-GPS-240 Section 3.2.2.2],
c. verify each System Effectiveness Model (SEM) almanac message created by the GPS service provider meets the format and transmission requirements specified in ICD-GPS-240 [ICD-GPS-240 Section 3.2.2.3], and
d. verify each Yuma almanac message created by the GPS service provider meets the format and transmission requirements specified in ICD-GPS-240 [ICD-GPS-240 Section 3.2.2.3].

3.5 REPORTING AND NOTIFICATION REQUIREMENTS

a. All events shall be reported to the satellite operations as part of their normal operational duties[8]. To the extent practical, reports shall include the measured or calculated values, the threshold values that are exceeded, and shall identify the source of the data.
b. Civil monitoring shall provide electronic notification of events to agencies identified to receive notification through published documents and memorandums of agreement (currently U.S. Coast Guard and the Federal Aviation Administration) for further distribution to user groups as required and with a timeliness to which each side has agreed [ICD-GPS-240 Sections 10, 20, 30].
c. Civil monitoring shall detect events in the times specified in Table 3.5-1[9,10].

---

[8] Civil monitoring and responses to reported civil service and civil signal events need to be incorporated into the service provider's standard operating procedures in order to assure a timely response to events.

[9] These requirements were reviewed and accepted through a series of Signal Monitoring Working Group meetings held March 31-April 2, 2008, May 14-15, 2008, and October 28-29, 2008 as documented in the minutes of the April 20, 2009 meeting of the CMPS Review Committee.

[10] The values in Table 3-5.1 were selected to be commensurate with the type of the event and the monitoring interval.

GPS Civil Monitoring Performance Specification

Table 3.5-1 Event Detection Times

Event Title	Detection Time
Constellation management events (Section 3.1.1)	Within 24 hours of onset of event
Signal in space coverage events (Section 3.1.2)	Within 24 hours of onset of event
Signal in space accuracy events (Section 3.1.3)	Within 1 day after data collection period
Signal in space reliability events (Section 3.1.4)	Within 1 minute after data collection period
Signal in space continuity events (Section 3.1.5)	Within 1 day following transmission of erroneous status and problem report
Signal in space availability events (Section 3.1.6)	Within 24 hours of onset of event
Position/time domain availability events (Section 3.1.7)	Within 24 hours of onset of event
Position/time domain accuracy events (Section 3.1.8)	Within 1 day after data collection period
Civil ranging code events (Section 3.2.1)	Within 1 minute of onset of event
Civil signal absolute power (Section 3.2.2.a-f)	Within 90 days of onset of event
Civil signal relative power (Section 3.2.2.g-j)	Within 1 hour of onset of event
Civil signal deformation (Section 3.2.2.k-y)	Within 1 minute of onset of event
Semi-codeless tracking events (Section 3.2.3)	Within 5 minutes of onset of event
Navigation message events (Section 3.2.4)	Within 1 minute following transmission

d. Civil monitoring shall report current GPS service and signal availability and accuracy levels to the service provider and to civil interface agencies[9,11].
e. Limitations or failures in civil monitoring that restrict the ability to fulfill the requirements defined in 3.1 or 3.2 shall be reported[9].

3.6 ANALYSIS AND DATA ARCHIVING REQUIREMENTS

This section summarizes the data analysis and data archiving requirements necessary to support civil monitoring. In order to support civil monitoring, the organization(s) that perform civil monitoring shall:

a. retain copies of all raw sensor data for a period not less than seven years[9,12,13],
b. retain copies of all reports issued as a result of civil monitoring through the design life of the system[9,13,14],
c. retain the results of the analyses performed for a period not less than seven years[9,13],
d. assess the time to issue a NANU prior to a scheduled event as identified in Section 3.2 [SPS PS Section 3.6],

---

[11] This requirement ensures that the civil interface organizations will have information on the current performance of GPS, even when it is meeting or exceeding required performance.

[12] This requirement is intended to ensure sufficient time for the U.S. Government to resolve legal and international issues related to service provision. Data to be retained includes, but is not limited to, raw sensor observations, navigation message data, and all derivative products.

[13] For 3.6 a, b, and c, the intent is for all data to be retained within the monitoring system and not moved off-line. That is to say, the data should be accessible without requiring additional steps for access as it ages or increased delay over contemporaneous data.

[14] This requirement will ensure adequate record keeping for a government-provided service.

e. assess the time to issue a NANU following an unscheduled event as identified in Section 3.2 [SPS PS Section 3.6],
f. provide information identifying integrity failures[15] for inclusion in a GPS integrity anomaly database, including date/time of onset, duration, failure description, magnitude, and affected satellite (if applicable) [GPS IFMEA 2002 Final Report, Section VI], and
g. retrieve and display up to the past 30 days of measured and calculated monitoring results within 6 seconds of a request from operations [9,16].

3.7 INFRASTRUCTURE REQUIREMENTS

This section describes the requirements levied on the infrastructure to ensure the availability and usability of civil monitoring data.

a. The civil monitoring capability shall detect and reject raw measurement data that has been tampered with, and shall notify operations of such instances[9,17,18].
b. Civil monitoring shall collect the necessary observations to perform the analyses in Section 3.5 from all SVs continuously and with sufficient redundancy to support unambiguous isolation of errors[9,19].
c. The status of data collection, transmission, and analysis infrastructure that supports monitoring shall be monitored and events that reduce the data available in support of monitoring or degrade the quality of data available in support of monitoring shall be reported to the monitoring operations and recorded by the monitoring system[9].

3.8 OPERATIONS INTEGRATION REQUIREMENTS

a. The results provided or produced by civil monitoring shall be incorporated into satellite operations, including daily operation and standards and evaluation processes[9,20].
b. The monitoring function shall be available 99.9% over any 365 day period (approximately 8 hours of outage for one year). The monitoring function shall be maintained during routine deployment and routine maintenance, to include software updates[9,16].
c. Notification of GPS operations of trends approaching performance failures (SS-SYS-800 Section 3.2.3.2.1 and SPS PS 3.5) shall be provided within 7.5 minutes (threshold) with an objective of 2 minutes from system time of receipt of data containing failure indications[9,16].

---

[15] For purposes of this document, integrity failures are those failures identified in the 2002 IFMEA Report.

[16] This requirement comes from 2008 Signal Monitoring Working Group discussions with 19 SOPS officials.

[17] This requirement protects the integrity of the monitoring results by assuring the integrity of the source data. Exceptions to this requirement may be considered during system architecture definition and system design if such exceptions address tamper detection and strengthen the overall integrity of the system.

[18] This requirement comes from 2008 Signal Monitoring Working Group discussions with USAF officials, including AFSPC/A5, 14AF, and 2 SOPS.

[19] This requirement is derived from the detection time requirements stated in Table 3.5-1. In order to detect specified signal and navigation message events within a minute, it is necessary to have continuous observations (at least every Z-count). In order to have confidence in the detection, it is necessary to have redundant observations from at least two sites throughout normal operations (i.e. even during normal maintenance activities).

[20] Experience has demonstrated that activities not incorporated into the day-to-day process of satellite operations are not fully implemented and may be lost or ignored.

4 PARTITIONING OF REQUIREMENTS

Table 4-1 presents a partitioning of the requirements between those that are judged unique to the civil community and those that are believed to be more general in nature. Only the requirements in Sections 3.1, 3.2 and 3.4 are covered in Table 4-1. Requirements in the remaining sub-sections of Section 3 are requirements associated with the monitoring system itself and are applicable in either case.

Table 4-1 – Partitioning of Requirements

Section	Civil Unique	General
3.1.1.a-b		X
3.1.2.a-b		X
3.1.3.a-h		X
3.1.4.a-d[21]	X	X
3.1.5.a-c	X	
3.1.5.d		X
3.1.6.a-d		X
3.1.7.a-f	X	
3.1.8.a-e	X	
3.2.1.a		X
3.2.1.b	X	
3.2.1.c		X
3.2.1.d-e	X	
3.2.1 f		X
3.2.1.g-m	X	
3.2.1 n[21]	X	X
3.2.1.o		X
3.2.1.p-r	X	
3.2.2.a		X
3.2.2.b-h	X	
3.2.2.i		X
3.2.2.j-l	X	
3.2.2 m		X
3.2.2 n-p	X	
3.2.2.q		X
3.2.2 r-t	X	
3.2.2.u		X
3.2.2.v	X	
3.2.2.w		X
3.2.2.x-z	X	
3.2.2.aa		X
3.2.3.a-b	X	
3.2.3.c-d		X
3.2.4.1.a-y		X
3.2.4.2.a-w	X	
3.2.4.3.a-s	X	
3.2.4.4.a-t[21]	X	
3.4.a-d		X

---

[21] Civil monitoring of L1 C/A are general requirements while monitoring of L2C, L5, and L1C signals are civil unique requirements.

5 NOTES

5.1 ADDITIONAL REFERENCES

The following documents and articles provide useful reference information in addition to those documents listed in Section 2.

1. *Department of Defense World Geodetic System 1984, Its Definition and Relationships with Local Geodetic Systems*, DMA Publication TR-8350.2 (unlimited distribution), Second Edition, September 1, 1991.
2. Clyde R. Greenwalt and Melvin E. Shultz, *Principles of Error Theory and Cartographic Applications*, United States Air Force Aeronautical Chart and Information Center Publication ACIC Technical Report No. 96 (unlimited distribution), February 1962.
3. Gerald J. Hahn and William Q. Meeker, *Statistical Intervals: A Guide For Practitioners* (New York: John Wiley & Sons, Inc., a Wiley-Interscience Publication, 1991).

5.2 GPS CIVIL MONITORING SERVICE USE CASES

The following use cases illustrate the anticipated applications of civil monitoring.

5.2.1 GPS Operational Command and Control

This use case describes how civil monitoring is used to support GPS mission operations and ensure highest availability of service.

Concept of Operation: The USAF manages GPS through its 2nd Space Operations Squadron (2 SOPS) which maintains the health and status of the operational constellation at facilities located at Schriever Air Force Base, Colorado. The 2 SOPS provides round the clock assessment of GPS performance and periodic updates to the spacecraft to maintain an accurate and dependable service. These assessments are made using measurement and signal status data obtained from the world-wide network of monitor stations. If anomalies occur, that is, instances in which service is outside of established thresholds, the operators will take corrective action to mitigate impact to users. This could include setting a satellite unhealthy, shutting down a satellite subsystem, or performing a contingency upload.

Entry Criteria:
- GPS is operational

Actors:
- Satellite Operators (2 SOPS currently)
- Supporting monitoring facilities/operators
- GPS operational control segment and space segment

Description:
- Civil GPS service is monitored
- Service anomaly is detected
- Trends approaching Misleading Signal Information (MSI) events detected
- Satellite Operators are notified

	• Satellite Operators take action to remedy service anomaly
Exit Criteria:	• Service anomaly detected within time specified
	• Service anomaly remedied by satellite operators
Requirements Verified	• Sections 3.5a, 3.8

5.2.2 GPS Service Standard Adherence

This use case describes how civil monitoring is used to verify U.S. Government commitments to GPS users.

Concept of Operations: The U.S. Government has made commitments for service by the establishment of standards, plans, and other documents describing expected service levels. These commitments for service are verified by civil monitoring by computing metrics for actual services levels and comparing these to the thresholds set forth in the standards, plans, and other documents describing expected service levels.

Entry Criteria:	• GPS is operational
Actors:	• Satellite Operators (2 SOPS currently)
	• Supporting monitoring facilities/operators
	• GPS operational control segment and space segment
Description:	• Civil GPS service is monitored
	• Service anomaly is detected
	• Satellite Operators are notified
	• Appropriate civil agencies are notified (Section 3.5d)
	• Satellite Operators take action to remedy service standard failure
Exit Criteria:	• Service anomaly detected and resolved within time specified
Requirements Verified	• Sections 3.1.1, 3.1.2, 3.1.3, 3.1.4, 3.1.5, 3.1.6, 3.1.7, 3.1.8, 3.4, 3.5b, 3.5d

5.2.3 GPS SPS and IS Compliance

This use case describes how civil monitoring is used to verify the compliance of the GPS signal with U.S. Government specifications.

Concept of Operations: The U.S. Government has made commitments for signal performance by the establishment of standards, interface specifications, and other documents describing expected signal performance levels. These commitments for signal performance are verified by civil monitoring by computing metrics for actual signal performance levels and comparing these to the thresholds set forth in the standards, interface specifications and other documents describing expected performance levels.

Entry Criteria:	• GPS is operational
Actors:	• Satellite Operators (2 SOPS currently)
	• Supporting monitoring facilities/operators
	• GPS operational control segment and space segment
Description:	• Civil GPS service is monitored
	• Service anomaly is detected
	• Satellite Operators are notified
	• Satellite Operators take action to remedy signal specification failure
Exit Criteria:	• Service anomaly detected and resolved within time specified
Requirements Verified	• Sections 3.2.1, 3.2.2 (m-aa), 3.2.3, 3.2.4, 3.5a

5.2.4 Situational Awareness

This use case describes how civil monitoring is used to provide user interface organizations with a real-time and predicted situational awareness of GPS service.

Concept of Operation: The operators and various external organizations are very interested in knowing the GPS service being provided throughout their service areas. Under this use case, civil monitoring assesses and distributes the status of performance in the form of geographic informational data overlaid onto maps, tables of running statistics, and real-time status and measurement values. Such information is tailored to match individual areas of interest. Means of notification may include current controlled interfaces and distribution via secure and public channels.

Entry Criteria:	• GPS is operational
Actors:	• GPS operational control segment and space segment
Description:	• Civil GPS service is monitored
	• Civil monitor reports status of constellation to appropriate agencies (Section 3.5d)
Exit Criteria:	• Reports are created and distributed
Requirements Verified	• Sections 3.1.2-3.1.8, 3.5, and 3.6

GPS Civil Monitoring Performance Specification

5.2.5 Past Assessment

This use case describes how civil monitoring is used to assess past service at any time in any part of the world. Such a capability would be useful in resolving liability claims or misinformation regarding GPS performance.

Concept of Operation: There are times in which the U.S. Government is asked to report on or substantiate the levels of service provided by GPS in times past. Examples of this include inquiries in criminal litigation cases, from the National Transportation Safety Board, or from operational military teams doing battle damage assessment. By maintaining an easily accessible archive of GPS civil performance data, civil monitoring is able to readily meet these requests for data.

Entry Criteria:	• GPS is operational
Actors:	• GPS operational control segment and space segment
Description:	• Civil GPS service is monitored
	• Civil monitor records performance of civil GPS service
	• Civil monitor generates reports and analyses for past periods as requested by service provided and/or civil interface agencies
	• Civil monitor generates records of integrity failures for inclusion in a GPS integrity anomaly database
Exit Criteria:	• Reports are created and distributed
Requirements Verified	• Section 3.5 and 3.6

5.2.6 Infrastructure

This use case describes how civil monitoring uses sites outside the USAF and NGA networks for monitoring the civil signals.

Concept of Operation: The normal path for performance data has been USAF and NGA monitor stations providing satellite ranging and status data to the Master Control Data. In order to broaden the reach of the civil monitoring, other monitoring sites are included, such as those operated by U.S. Government agencies (e.g., NASA), and those operated by non-U.S. Government organizations (e.g., universities and foreign governments). If non-secure data is employed for civil monitoring, appropriate measures are taken to ensure that it has not been tampered with.

Entry Criteria:	• GPS is operational
	• GPS data is received from monitoring sites
Actors:	• GPS operational control segment and space segment

Description:	• U.S. Government GPS monitoring sites other than GPS control segment
	• Non-U.S. GPS monitoring sites
	• Civil GPS service is monitored
	• Civil monitor records performance of civil GPS service
	• Civil monitor detects, rejects, and reports measurements that have been tampered with
	• Civil monitor collects and reports status data of monitoring networks to monitoring operators
Exit Criteria:	• Cases of tampered data are detected, rejected, and reported
	• Anomalies are isolated unambiguously
	• Monitoring network status data are reported
Requirements Verified	• Section 3.7

5.2.7 Power Level Assessment

This use case describes how civil monitoring is used to detect degradation of user received signal power either due to natural degradation of components over time or human control of signal power.

Concept of Operation: For the user, received, not transmitted, signal power is important. Natural degradation of signal power is assessed periodically (annually) by an organization having a directional receiving antenna with a known gain and noise level that is able to measure received power from a given satellite. This is done for each satellite. This assessment is also performed when requested by operators, typically when they suspect a satellite is performing below specification and needs to be checked. To provide continuous monitoring between the periodic checks of the received signal power, the C/N_0 is assessed at each monitoring receiver. This is done for each satellite, and the measurements are combined and filtered to generate an aggregate power level value for each satellite. The aggregate values are then examined for significant and unexpected drops in power level. In some cases, the operators may then request a received signal power check to be performed using a directional antenna.

Entry Criteria:	• GPS is operational
Actors:	• Satellite Operators (2 SOPS currently)
	• Signal power monitoring facility (such as Camp Parks)
	• Supporting monitoring facilities/operators
	• GPS operational control segment and space segment
Description:	• Civil GPS signal received power is assessed periodically and on demand

- Civil GPS signal C/N_0 is monitored continuously
- Degraded power is observed
- Satellite Operators are notified
- Satellite Operators take action to remedy degraded signal power

Exit Criteria: Requirements Verified
- Signal power level anomaly resolved within time specified
- Sections 3.2.2.a-l

5.3 ALLOCATION OF REQUIREMENTS TO USE CASES

This section provides an allocation of the requirements in Section 3 to the use cases in Section 5.2.

Table 5.3-1 Allocation of Requirements to Use Cases

Requirement	GPS Operational Command and Control	GPS Service Standard Adherence	GPS SPS and IS Compliance	Situational Awareness	Past Assessment	Infrastructure	Power Level Assessment
3.1.1 Verification of Constellation Management Standard		X					
3.1.2 Verification of Space Coverage Standard		X		X			
3.1.3 Verification of Space Accuracy Standard		X		X			
3.1.4 Verification of Signal in Space Reliability Standard		X		X			
3.1.5 Verification of Signal in Space Continuity Standard		X		X			
3.1.6 Verification of Signal in Space Availability Standard		X		X			
3.1.7 Verification of Position/Time Domain Availability		X		X			
3.1.8 Verification of Position/Time Domain Accuracy		X		X			
3.2.1 Verification of Civil Ranging Codes			X				
3.2.2 Civil Signal Quality Monitoring			X				X
3.2.3 Verification of Signal for Semi-Codeless Tracking			X				
3.2.4 Verification of Navigation Message			X				
3.4 Non-Broadcast Data Monitoring Requirements		X					
3.5 Reporting and Notification Requirements	X	X	X	X	X		
3.6 Analysis and Data Archiving Requirements				X	X		
3.7 Infrastructure Requirements						X	
3.8 Operations Integration Requirements	X						

5.4 CLARIFICATIONS AND ALGORITHMS

5.4.1 Verification of Absolute Power

The goal of the requirements in Section 3.2.2 a-h is periodic (e.g. at least yearly) verification of the absolute power delivered by each SV. The operational organization shall also have the ability to request such verification, but the response time may be days or weeks before the verification is performed. Reference the use case in Section 5.2.7.

5.4.2 Received Carrier to Noise

These requirements address the fact that absolute signal power is difficult to obtain on a real-time continuous basis. However, a change in the absolute power will be reflected in a drop in the C/N_0 values. Assuming a real-time continuous monitoring capability exists, such a system can continuously (at least at a rate of once every 1.5 s) monitor the C/N_0 values provided by all stations tracking each SV. It is recognized that C/N_0 are inherently noisy and that there exists a dependency between C/N_0 and elevation angle. Therefore an actual system implementation will likely incorporate a variety of features such as smoothing over time, comparisons against historical values, normalization over a range of elevation angles. Whatever the implementation, the goal is to detect unanticipated discontinuities in C/N_0. Such a discontinuity will trigger a detailed examination using additional types of monitor station data which, if substantiated, may result in a request for a measurement of absolute power (see 5.4.1)

5.4.3 Code-Carrier Divergence and Code-Carrier Divergence Failure

The requirements in Section 3.2.2 k-n address code-carrier divergence. Code-carrier divergence at frequency Li calculated between times t and $t+T$ and using the carrier ranges of signals Lj and Lk for ionospheric correction is defined as

$$CCD_{Lj,Lk}^{Li}(t,t+T) = PR_{Li}(t+T) - PR_{Li}(t) - \left[CR_{Li}(t+T) - CR_{Li}(t) \right] - 2\left(\frac{f_{L1}}{f_{Li}}\right)^2 \Delta I_{Lj,Lk}(t,t+T)$$

where $\Delta I_{Lj,Lk}(t,t+T)$ is the difference in the $L1$ ionospheric delay between times t and $t+T$ calculated from dual Lj and Lk carrier range differences; i.e.,

$$\Delta I_{Lj,Lk}(t,t+T) = \frac{CR_{Lj}(t+T) - CR_{Lj}(t) - \left[CR_{Lk}(t+T) - CR_{Lk}(t) \right]}{1 - \left(\frac{f_{Lj}}{f_{Lk}}\right)^2}$$

With $CCD_{Lj,Lk}^{Li}(t,t+T)$ defined as above and for a GPS satellite track observed/monitored from times t_1 to t_2, a code-carrier divergence failure is defined to exist at time $t+T$ if all of the following conditions are satisfied.

$$100 \leq T \leq 7200$$
$$t_1 \leq t \leq t_2 - T$$

$$CCD_{Lj,Lk}^{Li}(t, t+T) > 6.1m$$

Parameters in the above equations are defined as follow:
t = time in seconds, $t = 1,2,3$ ---,
$Li \in \{L1, L1C, L2, L5\}$ (i.e., denotes signal type, either L1 (L1-C/A), L1C, L2 (L2C) or L5)
$Lj \in \{L1, L1C\}$ (i.e., Lj denotes either L1 (L1-C/A) or L1C)
$Lk \in \{L2, L5\}$ (i.e., Lk denotes either L2 (L2C) or L5)
$PR_{Li}(t)$ = Li pseudorange at time t (same for $PR_{Lj}(t)$ and $PR_{Lk}(t)$)
$CR_{Li}(t)$ = Li carrier range at time t
f_{Li} = Li frequency in Hz

The CCD requirement is satisfied for a given civil signal type if <u>any</u> of the four possible calculations for that signal type (see Table 5.4-1) complies.

Table 5.4-1 - Possible Combinations of Observables for CCD Calculations*

Coherency Check for	Possible Observables							
	PR_{L1}	CR_{L1}	PR_{L1C}	CR_{L1C}	PR_{L2}	CR_{L2}	PR_{L5}	CR_{L5}
L1	1.0	2.09				-3.09		
	1.0	-1.0		3.09		-3.09		
	1.0	1.52						-2.52
	1.0	-1.0		2.52				-2.52
L1C		3.09	1.0	-1.0		-3.09		
			1.0	2.09		-3.09		
		2.52	1.0	-1.0				-2.52
			1.0	1.52				-2.52
L2		5.09			1.0	-6.09		
		4.15			1.0	-1.0		-4.15
					5.09	1.0	-6.09	
					4.15	1.0	-1.0	-4.15
L5		5.54				-5.54	1.0	-1.0
		4.52					1.0	-5.52
					5.54	-5.54	1.0	-1.0
					4.52		1.0	-5.52

* Combinations with ionospheric corrections calculated via L2 and L5 frequencies not included

5.4.4 Signal Distortion

Without bounding the digital distortion of individual signals, the pseudorange biases estimated by the control segment (or differential reference station) and end user's equipment could diverge. At the time of the writing of this edition of this CMPS, the U.S. Government had not made commitments regarding the degree of uniformity in signal formation. Neither the SPS PS nor the interface specifications spoke to the issue of signal distortion. These are addressed to some degree in the SARPs, but only for L1 C/A code. A paper written by Dr. Christopher Hegarty and Dr. A.J. Van Dierendonck assessed what levels of chip lead and lag were allowed for the L1

C/A, L2C, L5I and L5Q, and L1C signals. (See C. Hegarty, A. Van Dierendonck, "Recommendations on Digital Distortion Requirements for the Civil GPS Signals", IEEE/ION PLANS 2008, May 6-8, 2008, Monterey, CA) Since monitoring of signal distortion has been deemed critical by the Signal Monitoring Working Group of 2008, this version of this CMPS takes the results of the work done by Hegarty and Van Dierendonck and applies it to the civil monitoring function described in Section 3.2.2, Civil Signal Quality Monitoring.

Even though the SARPs permit up to 0.12 chips in error magnitude in lead/lag, Hegarty and Van Dierendonck say that 0.02 L1C/A and L2C chips (19.6 ns) is the maximum asymmetry permitted to meet a -40dBc attenuation requirement for spurious emissions. L5I and L5Q are even tighter with a 0.02 chip (1.96 ns) maximum asymmetry permitted. L1C has a maximum asymmetry of 0.05 chips (6.2 ns).

5.4.5 Carrier Phase and Bit Monitoring

Anomalies have been observed on multiple GPS satellites relating to carrier phase discontinuities that then can introduce subframe parity errors due to navigation message bit inversions. Appropriate signal monitoring can be used to detect and isolate these behaviors in order to protect users. For example, in today's WAAS, carrier phase discontinuities are detected and individual satellite signals are flagged as "Not monitored", meaning they are not to be used. For effective identification and isolation of carrier phase discontinuities and bit inversions, a monitoring system must have multiple observations of each signal.

5.3.5.1 Carrier Phase Discontinuities

Carrier phase discontinuities have been observed in GPS satellite signals. The extent of these discontinuities varies from tens to hundreds of milliseconds. To most receivers, these phase discontinuities are manifested as cycle slips, either as partial cycle, half cycle or full cycle slips. While there is no explicit provision in the interface specification restricting such phase discontinuities, under the basic definition of the signal structure such behavior constitutes an anomaly. Such carrier phase discontinuities are not inherent in a properly formed signal and are anomalous to users. Online signal monitoring is needed to detect these aberrations for the purpose of mitigating them and protecting users.

Isolation of carrier phase discontinuities may require a special class of GPS equipment and introduce offline monitoring requirements to sufficiently characterize the anomalous performance to support fault identification. Since the characteristics of the events observed to date remain under investigation, it is premature to specify what is required to monitor and then fully characterize these events. Once the anomaly investigation is completed, there will be further delineation of monitoring details. However, monitoring for these occurrences will require the ability to correlate data from multiple receivers tracking the same signal to confirm that the aberration is signal dependent and not the result of atmospheric or local phenomena.

5.3.5.2 Bit Inversion

When a carrier phase discontinuity occurs, this may result in a bit inversion in the receiver demodulation of the navigation message. If this occurs, a parity failure will be observed using the CRC parity bits in the navigation message. Signal monitoring algorithms therefore should: (1) detect parity failures at each receiver processing a given satellite's signal; and (2) compare these parity failure results with each of the monitor receivers tracking the same signal. If the failure mode is the same for each receiver, that is, each receiver observes parity failure, then it is a strong indication that the satellite signal is at fault, and the fault is not due to local or atmospheric phenomena. This type of monitoring must be done on each of the navigation message types: L1C/A, L2C, L5I, and L1C.

5.4.6 Assessment of DOP Availability

The availability-of-DOP metric is defined in the following steps. This describes a general computation of DOP availabilities for sites and grids.

STEP 1. Define the performance assessment interval and sample rate, and the location or area within the service volume to be evaluated. For area assessments, use the equidistant spacing algorithm defined in Section 5.4.6.1 to identify the area boundaries and specific discrete locations within the area to be evaluated. The specified region can be of any size up to and including the entire globe.

STEP 2. Establish the specific type and magnitude of DOP thresholds required, and compute availability of DOP values over the assessment interval for each site within the specified grid. DOP values are computed using standard algorithms such as those described in *Global Positioning System: Signals, Measurements and Performance, Misra and Enge, 2nd Edition*, page 208.

In the algorithm below, the quantity "n" is the counter defining the number of samples over the performance assessment interval. If the "increment time" is set to 60 seconds and the performance assessment interval is 24 hours; the value of "n" is 1,440. The quantity "m" is the number of points in a grid run. If "increment degrees" is set to 10 degrees for example, the value of "m" is 468. The counters "i" and "j" are used to indicate the sequential time step for each point and the grid point within the grid sequence, respectively. The quantities "HDOP_COUNT", "VDOP_COUNT", "PDOP_COUNT" and "TDOP_COUNT" are simple counters that are incremented each time the predicted errors are at or below the established thresholds. The quantity "NONAV_COUNT" is incremented whenever less than four satellites are available.

```
FOR i = 1 to n
IF HDOP ≤ HDOP_THRESHOLD THEN HOR_COUNT = HDOP_COUNT +1
IF VDOP ≤ VDOP_THRESHOLD THEN VERT_COUNT = VDOP_COUNT +1
IF PDOP ≤ PDOP_THRESHOLD THEN POS_COUNT = PDOP_COUNT +1
IF TDOP ≤ TDOP_THRESHOLD THEN POS_COUNT = TDOP_COUNT +1
```

$$AVAIL_{site}^{hdop} = \frac{\left(\sum_{i=1}^{n} HDOP\_COUNT_i - \sum_{i=1}^{n} NONAV\_COUNT_i\right)}{n}$$

$$AVAIL_{site}^{vdop} = \frac{\left(\sum_{i=1}^{n} VDOP\_COUNT_i - \sum_{i=1}^{n} NONAV\_COUNT_i\right)}{n}$$

$$AVAIL_{site}^{pdop} = \frac{\left(\sum_{i=1}^{n} PDOP\_COUNT_i - \sum_{i=1}^{n} NONAV\_COUNT_i\right)}{n}$$

$$AVAIL_{site}^{tdop} = \frac{\left(\sum_{i=1}^{n} TDOP\_COUNT_i - \sum_{i=1}^{n} NONAV\_COUNT_i\right)}{n}$$

STEP 3. Compute availability-of-DOP values over the assessment interval for the specified grid.

$$AVAIL_{grid}^{hdop} = \frac{\left(\sum_{j=1}^{m}\sum_{i=1}^{n} HDOP\_COUNT_{ij} - \sum_{j=1}^{m}\sum_{i=1}^{n} NONAV\_COUNT_{ij}\right)}{m \times n}$$

$$AVAIL_{grid}^{vdop} = \frac{\left(\sum_{j=1}^{m}\sum_{i=1}^{n} VDOP\_COUNT_{ij} - \sum_{j=1}^{m}\sum_{i=1}^{n} NONAV\_COUNT_{ij}\right)}{m \times n}$$

$$AVAIL_{grid}^{pdop} = \frac{\left(\sum_{j=1}^{m}\sum_{i=1}^{n} PDOP\_COUNT_{ij} - \sum_{j=1}^{m}\sum_{i=1}^{n} NONAV\_COUNT_{ij}\right)}{m \times n}$$

$$AVAIL_{grid}^{tdop} = \frac{\left(\sum_{j=1}^{m}\sum_{i=1}^{n} TDOP\_COUNT_{ij} - \sum_{j=1}^{m}\sum_{i=1}^{n} NONAV\_COUNT_{ij}\right)}{m \times n}$$

5.4.6.1 The Equidistant Spacing Algorithm

The objective of this algorithm is to generate the latitude and longitude of a sequence of points equal distances apart for all or a specified portion of the globe, given an input of the start and stop points and the desired distance between points. This algorithm is used to generate points for performing geometry computations or position solutions across any desired area at any required discrete density. The reason for using this algorithm is to ensure an even distribution of points over the assessment area. A conventional latitude/longitude degree increment weights a performance assessment erroneously towards the higher latitudes. Note that the algorithm does generate small latitude residuals at the prime meridian that slightly distort the equal spacing need. The size of the residual grows directly as a function of the grid spacing. At a 1°grid spacing, the maximum latitude residual at any given longitude results in a deviation of less than 12 kilometers in the nominal distance between the first and last points.

In this algorithm, latitude increments from 0° to 90° north of the equator, and 0° to -90° south of the equator. Longitude begins at 0° at the Greenwich Meridian, and increments to 360° counterclockwise as viewed from the North Pole.

STEP 1. Define grid spacing.

This value represents the distance to use between points in the grid. Input can be defined either in terms of degrees or kilometers. At the equator, 111.1395 kilometers equals 1°. Note that if this input is specified in terms of degrees, the number of degrees requested will only apply at the equator. This is due to the fact that the number of kilometers per degree longitude decreases as latitude increases. At 80° latitude, 1° equals approximately 19 kilometers. The convention of "degrees" is used for this implementation.

INITIALIZE_INCREMENT_DEGREES = _____ (Degrees)
INCREMENT_KM = 111.3195 × INITIALIZE_INCREMENT_DEGREES

STEP 2. Define start and stop points in degrees latitude and longitude.

Note that the algorithm will use the starting longitude as the reference, and return to it for the next latitude increment. The algorithm is intended to increment in a northeasterly direction. The end longitude should be larger than the start longitude. To ensure this, add 360° to the end longitude. The end latitude should be larger than the start latitude.

START_LAT = _____ (Degrees)   START_LONG = _____ (Degrees)
END_LAT = _____ (Degrees)   END_LONG = _____ (Degrees)

$j = 0$
$k = 0$
LONGITUDE(j=0) = START_LONG
LATITUDE(k=0) = START_LAT

STEP 3. Perform geometry or position solution computations at starting point, and for each {j,k} increment.

From this algorithm's perspective, it doesn't matter if a single solution is performed at this point before incrementing to the next, or if all solutions over the specified time interval are computed.

STEP 4. Compute the number of longitude increments required at the current latitude.

The equatorial radius of the Earth (r) equals 6378.137 kilometers.

$$INCREMENT\_DEGREES(k) = \frac{360}{2\pi r \cos LATITUDE(k)} \times INCREMENT\_KM$$

$$LONG\_INCREMENT\_NUMBER(k) = INTEGER\left[\frac{END\_LONG - START\_LONG}{INCREMENT\_DEGREES(k)}\right]$$

STEP 5. Increment longitude by the LONG_INCREMENT_NUMBER value.

If the current increment exceeds the count, reset the longitude to START_LONG, and increment the latitude (STEP 6).

j = j+1
If j < LONG_INCREMENT_NUMBER (k)
LONGITUDE (j=j+1) = [LONGITUDE (j) + INCREMENT_DEGREES(k)]mod 360

If j ≥ LONG_INCREMENT_NUMBER(k), THEN
j = 0
LONGITUDE(j=0) = START_LONG

STEP 6. Compute latitude step size in degrees, and the latitude count.

If the latitude count is exceeded, the process is complete and the entire grid has been computed. Note that this algorithm begins with the lowest latitude, and works to the greater latitude. If the global case is being evaluated, use (-90°+ LAT_INCREMENT_DEGREES) as the latitude start point and (90°- LAT_INCREMENT_DEGREES) as the latitude end point.

LAT_INCREMENT_DEGREES = INITIALIZE_INCREMENT_DEGREES

$$LAT\_INCREMENT\_NUMBER = INTEGER\left[\frac{END\_LAT - START\_LAT}{LAT\_INCREMENT\_DEGREES}\right]$$

k=k+1
If k < LAT_INCREMENT_NUMBER
LATITUDE(k+1) = LATITUDE(k) + LAT_INCREMENT_DEGREES

If k ≥ LAT_INCREMENT_NUMBER STOP

5.4.7 Position/Time Domain Accuracy

5.4.7.1 Position Domain Accuracy

The measured position domain accuracy metric is defined in the following steps. This describes a general computation of position and time accuracies for sites and grids.

STEP 1. Define the performance assessment interval and sample rate, and the location or area within the service volume to be evaluated. For area assessments, use the equidistant spacing algorithm defined in Section 5.4.6.1 to identify the area boundaries and specific discrete locations within the area to be evaluated. The specified region can be of any size up to and including the entire globe.

STEP 2. Define specific environmental and physical environment constraints applicable to the instantaneous position error measurement conditions.

STEP 3. Compute instantaneous position error values as defined in Section 5.4.7.1.1 for all points in the specified grid over the performance assessment interval.

STEP 4. Take the absolute value of each estimate (in the case of vertical error), rank order the values, and find the n_{th} sample associated with the 95th percentile. S_{ACC} equals the number of samples over the measurement interval.

$\Delta \text{HOR95\_SITE} = \Delta h_{sis}$ value at $n = \text{INTEGER}(0.95 \times S_{ACC})$

$\Delta \text{VERT95\_SITE} = \Delta u_{sis}$ value at $n = \text{INTEGER}(0.95 \times S_{ACC})$

$\Delta \text{POS95\_SITE} = \Delta p_{sis}$ value at $n = \text{INTEGER}(0.95 \times S_{ACC})$

STEP 5. Sort the 95% values across the regional grid to determine the maximum horizontal and vertical values, to support a worst site assessment.

STEP 6. Compute the regional median 95% horizontal, vertical and position errors, to support a regional accuracy assessment.

$\Delta \text{HOR95\_REGION} = \Delta \text{HOR95\_SITE}$ value at $n = \text{INTEGER}(0.5 \times \# \text{ Grid Points})$

$\Delta \text{VERT95\_REGION} = \Delta \text{VERT95\_SITE}$ value at $n = \text{INTEGER}(0.5 \times \# \text{ Grid Points})$

$\Delta \text{POS95\_REGION} = \Delta \text{POS95\_SITE}$ value at $n = \text{INTEGER}(0.5 \times \# \text{ Grid Points})$

5.4.7.1.1 Instantaneous Position Accuracy

This section defines the specific process for computing instantaneous position solution error vectors.

The performance standards are based upon the mapping of instantaneous SIS UREs into a user position error vector through the linearized position solution, for the series of points comprising the performance assessment global grid.

STEP 1. Compute SIS URE values for all satellites visible at the given time above the specified mask angle.

STEP 2. Compute the position solution geometry matrix (**G**), and rotate it into local coordinates. The **G**-matrix (defined below) is composed of *n* row vectors, one for each of *n* satellites in view. Each row vector contains the x, y, z and time coordinate direction cosines associated with one of the satellite-to-user vector geometries, as they are defined in the WGS-84 ECEF coordinate system. The IS-GPS-200D equations (Table 20-IV) must be applied to determine instantaneous satellite position vectors at the time-of-transmission based upon the navigation message ephemeris or almanac.

$$G_{ecef}^{allSVs} = \begin{bmatrix} \frac{x_{site} - x_{sv1}}{R_{sv1}} & \frac{y_{site} - y_{sv1}}{R_{sv1}} & \frac{z_{site} - z_{sv1}}{R_{sv1}} & 1 \\ \bullet & \bullet & \bullet & \bullet \\ \bullet & \bullet & \bullet & \bullet \\ \frac{x_{site} - x_{svn}}{R_{svn}} & \frac{y_{site} - y_{svn}}{R_{svn}} & \frac{z_{site} - z_{svn}}{R_{svn}} & 1 \end{bmatrix} = \begin{bmatrix} G_x^{sv1} & G_y^{sv1} & G_z^{sv1} & 1 \\ \bullet & \bullet & \bullet & \bullet \\ \bullet & \bullet & \bullet & \bullet \\ G_x^{svn} & G_y^{svn} & G_z^{svn} & 1 \end{bmatrix}$$

where: $\{x_{site}, y_{site}, z_{site}\}$ = Station location in Cartesian coordinates

$\{x_{svj}, y_{svj}, z_{svj}\}$ = j_{th} satellite position coordinates at time-of-transmission based upon navigation message contents

R_{svj} = Estimated range from site to j_{th} satellite

Use the coordinate rotation matrix **S** to rotate the geometry matrix into local (East-North-Up, or ENU) coordinates. Local horizontal is defined to be the plane formed by the East-North axes. Local vertical is defined to be coincident with the Up axis. The **S**-matrix is defined below.

$$S = \begin{bmatrix} -\sin \lambda_{site} & \cos \lambda_{site} & 0 & 0 \\ -\sin \phi_{site} \cos \lambda_{site} & -\sin \phi_{site} \sin \lambda_{site} & \cos \phi_{site} & 0 \\ \cos \phi_{site} \cos \lambda_{site} & \cos \phi_{site} \sin \lambda_{site} & \sin \phi_{site} & 0 \\ 0 & 0 & 0 & 1 \end{bmatrix}$$

where: $\{\varphi_{site}, \lambda_{site}\}$ = Site latitude and longitude in local coordinates

The geometry matrix rotation is defined below. The result of the rotation is a geometry matrix defined with respect to local coordinate axes.

GPS Civil Monitoring Performance Specification

$$G_{enu}^{allSVs} = \left[S_{ecef \to enu} \times G_{ecef}^{allsvs,T} \right]^T$$

STEP 3. Compute the inverse direction cosine matrix (**K**) for each time t_k. The pseudo inverse equation can be used in a full rank linear system to gain satisfactory results for purposes of performance monitoring and assessment.

$$K = G = [G^T G]^{-1} G^T$$

STEP 4. Compute the SIS instantaneous position error vector ($\Delta \mathbf{x}_{sis}$) for each time t_k.

$$\Delta \bar{x}_{sis}(Site_m, t_k) = \left[G_{enu}^{allsvs} \right]^+ \Delta \bar{r}_{sis}(Site_m, t_k) = K_{enu}^{allsvs} \Delta \bar{r}_{sis}(Site_m, t_k), \text{ or}$$

$$\begin{bmatrix} \Delta e_{sis}(Site_m,t_k) \\ \Delta n_{sis}(Site_m,t_k) \\ \Delta u_{sis}(Site_m,t_k) \\ \Delta t_{sis}(Site_m,t_k) \end{bmatrix} = \begin{bmatrix} K_{11} & \bullet & \bullet & K_{1n} \\ K_{21} & \bullet & \bullet & K_{2n} \\ K_{31} & \bullet & \bullet & K_{3n} \\ K_{41} & \bullet & \bullet & K_{4n} \end{bmatrix} \begin{bmatrix} ERD(SV_1, Site_m, t_k) \\ \bullet \\ \bullet \\ ERD(SV_n, Site_m, t_k) \end{bmatrix} \quad \text{(meters)}$$

where: $\Delta \bar{x}_{sis}(Site_m, t_k)$ = SIS position solution error vector in local coordinates (east, north, up and time) at the k_{th} solution time for the m_{th} site
$\Delta \bar{r}_{sis}(Site_m, t_k)$ = ERD(SV$_j$,Site$_m$,tk) values from Step 1, for all satellites used in the k_{th} position solution at the m_{th} site

STEP 5. Compute the instantaneous SIS horizontal error (Δh_{sis}), SIS 3D position error (Δp_{sis}) and dynamic time transfer user error relative to USNO UTC ($\Delta t_{sis\_gps\text{-}usno}$) for each time t_k. The quantity ($\Delta t_{gps\text{-}usno}$) is a value provided daily by the USNO to the 2nd Space Operations Squadron (2 SOPS).

$$\Delta h(Site_m, t_k) = \left[(\Delta e(Site_m, t_k))^2 + (\Delta n(Site_m, t_k))^2 \right]^{\frac{1}{2}} \quad \text{(meters)}$$

$$\Delta p(Site_m, t_k) = \left[(\Delta e(Site_m, t_k))^2 + (\Delta n(Site_m, t_k))^2 + (\Delta u(Site_m, t_k))^2 \right]^{\frac{1}{2}} \quad \text{(meters)}$$

$$\Delta t_{sis\_gps-usno}(Site_m, t_k) = \frac{1 \times 10^9 \, ns/s}{c} * \Delta t_{sis}(Site_m, t_k) + \Delta t_{gps-usno}(Site_m, t_k) \quad \text{(nanoseconds)}$$

5.4.7.2 Time Domain Accuracy

The measured time transfer accuracy metric is defined in the following steps.

STEP 1. Define the performance assessment interval and sample rate, and the location or area within the service volume to be evaluated. For area assessments, use the equidistant spacing algorithm defined in Section 5.4.6.1 to identify the area boundaries and specific discrete locations within the area to be evaluated. The specified region can be of any size up to and including the entire globe.

STEP 2. Define specific environmental and physical environment constraints applicable to the instantaneous time transfer error measurement conditions.

STEP 3. Choose the specific time transfer algorithm to be used in the assessment. For static time transfer, use the algorithm below. For dynamic time transfer, use the time error output of the position solution algorithm defined in Section 5.4.7.1.1. Compute instantaneous time transfer error values for all points in the specified grid over the performance assessment interval.

Static Time Transfer User Solution
Users employing time transfer receivers from a surveyed location generally use an algorithm that is independent of the position solution geometry. Various sampling methods and smoothing intervals are used to generate the best possible estimate of their time offset relative to UTC(USNO). A conservative estimate of the error in determining a time transfer receiver's time scale offset to UTC(USNO) is provided below. The quantity (Δ $t_{gps-usno}$) represents the error in the knowledge of the bias between GPS time and UTC as it is defined by the USNO. The quantity "n" represents the number of satellites in view above the mask angle at time t_k.

$$\Delta t_{user}^{static}(Site_m, t_k) = \frac{1 \times 10^9 \, ns/s}{c * n} \sum_{j=1}^{n} ERD(SV_j, Site_m, t_k) \quad \text{(nanoseconds)}$$

STEP 4. Take the absolute value of each estimate (in the case of vertical error), rank order the values, and find the n_{th} sample associated with the 95th percentile. S_{ACC} equals the number of samples over the measurement interval.

Δ TIME95_SITE = Δ t_{user} value at n = INTEGER(0.95 x S_{ACC})

STEP 5. Sort the 95% values across the regional grid to determine the maximum time transfer error statistic, to support a worst site assessment.

STEP 6. Compute the regional median 95% time transfer error statistic, to support a regional accuracy assessment.

Δ TIME95_REGION = Δ TIME95_SITE value at n = INTEGER(0.5 x # Grid Points)

5.5 DEFINITIONS

The following definitions apply to the terms and acronyms used in this specification.

Accuracy	The degree of conformance between the estimated or measured position and/or velocity of a platform at a given time and its true position or velocity [Federal Radionavigation Plan (FRP), Appendix D]
Availability	The percentage of time that the services of a system are usable. Availability is an indication of the ability of the system to provide usable service within the specified coverage area. Signal availability is the percentage of time that navigation signals transmitted from external source are available for use. PDOP availability is the percentage of time over a specified time interval that the PDOP is less than or equal to a specified value [FRP, Appendix D]
Continuity	The ability of the total system to perform its function without interruption during the intended operation. The probability that the specified system performance will be maintained for the duration of a phase of operation, presuming that the system was available at the beginning of that phase of operation [FRP, Appendix D]
Coverage	The surface area or space volume in which the signals are adequate to permit the user to determine position to a specified level of accuracy [FRP, Appendix D]
Dilution of Precision	The magnifying effect on GPS position error induced by mapping URE into a position solution within the specified coordinate system, through the relative satellite-to-receiver geometry [SPS PS, Appendix C]
Healthy	The SPS SIS health is the status given by the real-time health-related information broadcast by each satellite as an integral SPS SIS. For further information, refer to SPS PS 2.3.2 [SPS PS 2.3.2]
Integrity	Integrity is a measure of the trust which can be placed in the correctness of the information supplied by the signal-in-space. Integrity includes the ability of the space and control segment to provide timely alerts or warnings (including switches to non-standard code) to users when the signal-in-space error may exceed the accuracy broadcast to the user [FRP, Appendix D]
Misleading Signal-In-Space Information	The pseudorange data set (e.g., raw pseudorange measurement and NAV data) provided by an SPS SIS provides Misleading Signal-in-Space information (MSI) when the instantaneous URE exceeds the SIS URE NTE tolerance [SPS PS, Appendix C].

Reliability	The probability of performing a specified function without failure under given conditions for a specified period of time [FRP, Appendix D]
Service volume	The spatial volume supported by SPS performance standards. Specifically, the SPS Performance Standard supports the terrestrial service volume. The terrestrial service volume covers from the surface of the Earth up to an altitude of 3,000 kilometers [SPS PS, Appendix C]
Signal availability	The percentage of time that navigation signals transmitted from an external source are available for use
Standard Positioning Service	The SPS is a positioning and timing service provided by way of ranging signals broadcast at the GPS L1 frequency. The L1 frequency, transmitted by all satellites, contains a coarse/acquisition (C/A) code ranging signal, with a navigation data message, that is available for peaceful, civil, commercial, and scientific use [SPS PS, Section 1.3]
User range error	The instantaneous difference between a ranging signal measurement (neglecting user clock bias), and the true range between the satellite and a GPS user at any point within the service

5.6 ABBREVIATIONS AND ACRONYMS

14AF	14th Air Force
2 SOPS	2nd Space Operations Squadron
19 SOPS	19th Space Operations Squadron
AFSPC	Air Force Space Command
AOD	Age of data
C/A	Course/Acquisition code
CL	Civil-long
CM	Civil-moderate
C/N_0	Carrier-to-Noise ratio
CMPS	Civil Monitoring Performance Specification
DoD	Department of Defense
DOT	Department of Transportation
EOP	Earth orientation parameters
FAA	Federal Aviation Administration

FRP	Federal Radionavigation Plan
GGTO	GPS/GNSS time offset
GIG	Global Information Grid
GPS	Global Positioning System
GPSW	Global Positioning Systems Wing
HOW	Handover word
ICD	Interface Control Document
IFMEA	Integrity, failure modes, and effects analysis
IFOR	Interagency Forum for Operational Requirements
IODC	Issue of data clock
IODE	Issue of data ephemeris
IS	Interface Specification
ITS	Intelligent Transportation System
LAAS	Local Area Augmentation System
MS	Monitor Station
MSI	Misleading Signal Information
NANU	Notice: Advisory to Navstar Users
NAV	Navigation message
NGA	National Geospatial-Intelligence Agency
NIS	Navigation Information Service
NOTAM	Notice to Airmen
NTE	Not to exceed
PDOP	Position dilution of precision
PPS	Precise Positioning Service
PRN	Pseudorandom noise
PRN ID	PRN Identifier
RF	Radio frequency
SARPs	Standards and Recommended Practices
SEM	System Effectiveness Model
SIS	Signal in space
SPS	Standard Positioning Service
SPS PS	SPS Performance Standard

SV	Space vehicle
TLM	Telemetry word
TOI	Time of interval
TOW	Time of week
URA	User range accuracy
URAE	User range acceleration error
URE	User range error
URRE	User range rate error
USAF	United States Air Force
USC	United States code
USCG	United States Coast Guard
UTC	Coordinated Universal Time
UTCOE	Coordinated Universal Time offset error
WAAS PS	Wide Area Augmentation Service Performance Standard
WN	Week number
WNt	Week number associated with leap second

www.ingramcontent.com/pod-product-compliance
Lightning Source LLC
Chambersburg PA
CBHW081907170526
45167CB00007B/3184